Playful Animals

CAPYBARAS

Ursula Pang

PowerKiDS press.

PK Beginners

Capybaras are rodents.

Rodents

Capybaras are
the biggest rodents!

Capybaras have brown fur.

Capybaras have
big teeth.

Capybaras eat plants.

Capybaras live in groups.

Capybaras like water.

Capybaras swim.

Capybaras are friendly.

Capybara babies are cute!

Capybaras are fun!

Published in 2025 by The Rosen Publishing Group, Inc.
2544 Clinton Street, Buffalo, NY 14224

First Edition

Editor: Greg Roza
Book Design: Michael Flynn

Photo Credits: Cover, p. 1 Steph Mattioli/Shutterstock.com; p. 3 (cayabara) Wirestock Creators/Shutterstock.com; p. 3 (mouse) Rudmer Zwerver/Shutterstock.com; p. 3 (rat) Heiko Kiera/Shutterstock.com; p. 3 (guinea pig) Miroslav Hlavko/Shutterstock.com; p. 5 wannawit_vck/Shutterstock.com; p. 7 Giedriius/Shutterstock.com; p. 9 Pertfoto/Shutterstock.com; p. 11 Foreverhappy/Shutterstock.com; p. 13 SandroSalomon/Shutterstock.com; p. 15 buteo/Shutterstock.com; p. 17 Marzio Gentili/Shutterstock.com; p. 19 Marcel_Strelow/Shutterstock.com; p. 21 Bruno Vieira/Shutterstock.com; p. 23 Isabela Nicoletti/Shutterstock.com.

Cataloging-in-Publication Data

Names: Pang, Ursula.
Title: Capybaras / Ursula Pang.
Description: Buffalo, NY : PowerKids Press, 2025. | Series: Playful animals
Identifiers: ISBN 9781499450675 (pbk.) | ISBN 9781499450682 (library bound) | ISBN 9781499450699 (ebook)
Subjects: LCSH: Capybara–Juvenile literature.
Classification: LCC QL737.R662 P36 2025 | DDC 599.35′9–dc23

Manufactured in the United States of America

CPSIA Compliance Information: Batch #CWPK25. For further information contact Rosen Publishing at 1-800-237-9932.